WHAT EXACTLY IS VIRTUAL REALITY?

What Exactly is Virtual Reality?

DAWSON MORTON

5d Media Publishing

© 2022 5D Media LLC. All rights reserved. This book or any portion thereof may not be reproduced or used in any matter whatsoever without the express written permission of the publisher except for the use of brief quotations in a book review. 5D Media Publishing division. contact@5dmedia.org

CONTENTS

What Exactly is Virtual Reality and the Metaverse?   1

The History of Virtual Reality   4

How Does it Work?   7

VR vs AR   12

How Is It Made?   16

Benefits & the Future of VR   19

LETTER FROM THE EDITOR

Virtual reality is a technology that has been around for years, but it is only now starting to become popular. Everyone should be knowledgeable about this technology as it is becoming more and more common for people to use virtual reality to entertain themselves. Educators and parents should provide students with information about virtual reality so that they can understand this technology and how it will shape the future. This knowledge will be very beneficial as technology improves, costs continue to go down, and virtual reality amplifies across all sectors of life, including entertainment, medicine, business, and education.

Student wearing VR headset
Kimberly Gordon

| one |

What Exactly is Virtual Reality and the Metaverse?

Virtual Reality (VR) is an immersive experience that can transport people to different worlds. It can be used for entertainment, education, and business purposes. Just about any experience or environment that can be imagined can be created in VR. As one of the most exciting technological developments of the 21st century, it is important to understand exactly what virtual reality is.

Artificial, computer-generated environments in VR allow users to interact with objects, sounds, and images from a simulated world. The metaverse is a term coined by Neal Stephenson, in his 1992 science fiction novel Snow Crash, to describe a universe of virtual reality worlds. It has been adopted by online virtual world communities to describe the collective of all such environments, both commercial and personal.

Man Paddling while wearing Virtual Reality Glasses
Rodnae

Virtual reality has been used in many areas, including gaming and training, for over 30 years. It was originally developed for military training simulations and has been used in medical and scientific research, including for treating phobias and post-traumatic stress disorder (PTSD).

Today, virtual reality is being used to create new forms of entertainment, as well as to create new forms of education and communication. VR experiences can simulate a physical presence in places in the real world or an imagined world, by enabling the user to interact with that world. The emphasis is on creating an illusion of reality. To achieve this, virtual reality immerses users into a totally artificial environment where they can explore and manipulate objects from a first-person perspective (using stereoscopic displays and motion tracking devices).

Girl using VR Goggles
Mikhail Nilov

Sensory experiences are also created, which can include sight, hearing, touch, smell and taste. The term "virtual reality" was first used in the 1960s to describe immersive simulations that immerse the user in an entirely synthetic environment. It has been described as a realistic and immersive simulation of an environment.

| two |

The History of Virtual Reality

Sensorama Simulator

Precisely when the first virtual reality (VR) device was created is not known, but it's thought that Morton Heilig invented a prototype head-mounted display called the Sensorama in 1957. The invention included stereoscopic goggles, earphones, 3D smell generators and fans to provide wind and aromas as well as motion seats for added realism.

It wasn't until 1968 that Ivan Sutherland developed what he called "the Sword of Damocles" at Harvard University, which was a head-mounted display connected to a computer and sensor gloves. His goal with this technology was to create an immersive experience in which users could manipulate objects in three dimensions on screen by moving their hands and arms through space.

In 1975 Myron Krueger introduced his version of VR into education with his electronic system named Videoplace, which allowed people living around the world to share an environment while they interacted with one another. He used avatars that were controlled by real actors wearing costumes that provided haptic feedback through sensors attached to them during filming sessions inside a studio set up like an Egyptian tomb or Roman palace. This film footage would then be broadcasted over regular television for other participants.

The technology was revolutionized by the introduction of computer-generated imagery (CGI) in 1982 with a program called Blue Marble created by artist Roy Trumbull. It was funded by NASA to showcase its vision of the Earth as a planet that could be harmed if humans didn't become responsible stewards of it.

The first VR headset was was made out of wood, then plastic, named "HTC Dream" or Tilt Brush. It started shipping on September 15, 2010 and was developed in collaboration between Google and HTC (and originally branded as a smartphone). It had two 1200×1920 LCDs screens with a pixel density that came out at 220 ppi with stereoscopic 3D support for viewing content such as photos, videos or rendered environments for games.

Later that year Sony announced Project Morpheus at GDC/Game Developers Conference around the same time Oculus VR revealed its prototype Rift Development Kit 2(DK2). That same month Zeiss unveiled special lenses designed specifically to be used inside headsets worn while playing video games so players could get more depth perception when looking into virtual worlds.

A Japanese company called PC Survice Co., Ltd would soon after release software designed to let you virtually experience the view from your head, it was called Google Cardboard. The Oculus

Rift DK2 went up for pre-order on January 6, 2014 after being announced in October 2013 and officially released at GDC in March 2014. It became available to purchase in February 2016.

Virtual reality technology has been around for decades, but it is only recently that it has become more mainstream and accessible to the average person. It was too expensive and cumbersome to be practical. Now, however, the cost of virtual reality has come down significantly and it is becoming increasingly popular. This is largely due to the advancements in mobile technology, which has allowed for the development of lightweight headsets that can be used with smartphones.

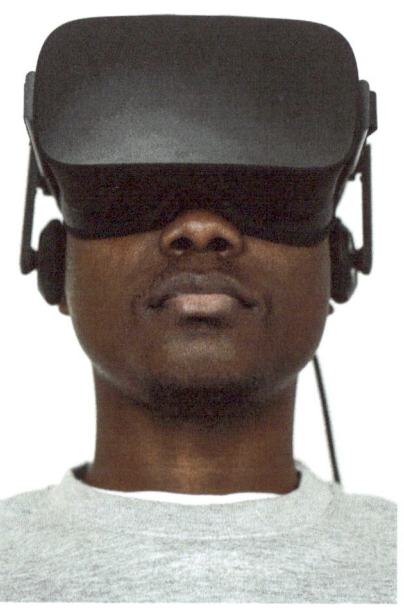

A Serious Man Wearing VR Headset
Ivan Samkov

| three |

How Does it Work?

Boy holding VR Headset
Julia M Camreron

Virtual reality works by first immersing you in a virtual world, then shutting off the real world. When everything around you is imaginary and your sense of touch is cut off from the outside, your mind starts to believe what it sees. The most advanced VR systems can fool your senses into believing that you are present in a different place. Virtual reality systems that include transmission of vibrations and other sensations to the user through a game controller or other devices are known as haptic systems. This tactile information is generally known as force feedback in medical, video gaming and military training applications. Some haptic systems include tactile sensors that can measure the force and velocity of applied forces, which can be used to control or simulate the force feedback.

The most common type of virtual reality system today is the head-mounted display (HMD). The HMD consists of goggles containing two small LCD screens that display images of the game you are playing or the experience you're viewing. Two small speakers in the headset deliver audio to match the video you see on your screen. Finally, a head tracking device is used to keep the image from your goggles fixed in place as you move around, giving you an illusion of being present inside the world of your game or application.

The problem with HMDs is that they limit what you can see since there's no way to turn your head far enough without bumping into something and breaking immersion (the feelings of being "present" inside a virtual world). That's where full-body VR comes in. Full body VR systems allow users to move their whole bodies while wearing motion capture sensors that track each movement digitally so they can be translated into movements within a virtual environment.

Ballerina dancing in VR
Rodnae

One such system was developed by researchers at Dartmouth University using off-the-shelf technology like Microsoft Kinect sensors and Oculus Rift goggles combined with open source software called OpenSimulator. This combination allowed for "full-body presence" where you feel like you're somewhere else entirely because it feels like part of your body is actually there! You don't just look around; now all parts of you are are able to explore a new world.

Full-body VR prototypes use existing game controllers, and the sensors are placed around a room or inside virtual reality gaming pods such as the Virtuix Omni platform. Because of this, full-body VR is mainly used for entertainment applications that allow you to play games while your whole body moves around (no limits to what you can do in real life!)--or be someone else entirely... like Spider-man! Most people do not have access to 8high tech motion capture suits or off-the-shelf hardware and software yet. These systems aren't cheap; not everyone will want them any time soon.

Hip Hop Gamer interview in the Metaverse
5d Media

| four |

VR vs AR

Virtual reality shares some elements with "augmented reality" (or AR) which is a type of virtual reality technology that blends what the user sees in their real surroundings with digital content generated by computer software. The additional software-generated images with the virtual scene typically enhance how the real surroundings look in some way.

Some augmentative virtual realities blend physical objects, such as toys or models, with virtual elements; others generate entirely new imaginary worlds. Augmented reality may also refer to non-immersive visual perception such as through smartphone cameras and display screens that allow information to be overlaid on top of an image being seen through them.

Close up of Augmented Reality through cell phone

Augmented reality is an overlay of digital information on top of the physical world, while virtual reality completely immerses users in a digital environment. One day soon, they will be indistinguishable and used interchangeably.

In 1992, Nicole Stenger created Angels, the first real-time interactive immersive movie where the interaction was facilitated with a dataglove and high-resolution goggles.

That same year, Louis Rosenberg created the virtual fixtures system at the U.S. Air Force's Armstrong Labs using a full upper-body exoskeleton, enabling a physically realistic mixed reality in 3D.

Girl playing in Virtual Reality
Mart

The system enabled the overlay of physically real 3D virtual objects registered with a user's direct view of the real world, producing the first true augmented reality experience enabling sight, sound, and touch.

Woman in Augmented Reality
Ali Pazani

| **five** |

How Is It Made?

To create virtual reality, you first need to understand what it is from a creator's perspective. Technically, virtual reality is a computer-generated simulation of a three-dimensional image or environment that can be interacted with in a seemingly real or physical way by a person using special electronic equipment, such as a helmet with a screen inside or gloves fitted with sensors.

The most common VR content to create is for a VR headset. This allows people to immerse themselves in the VR world. Another way to create VR content is with 360-degree video. This type of video gives people a 360-degree view of the world, and can be viewed without a headset. Finally, VR can also be created with 3D images and animations.

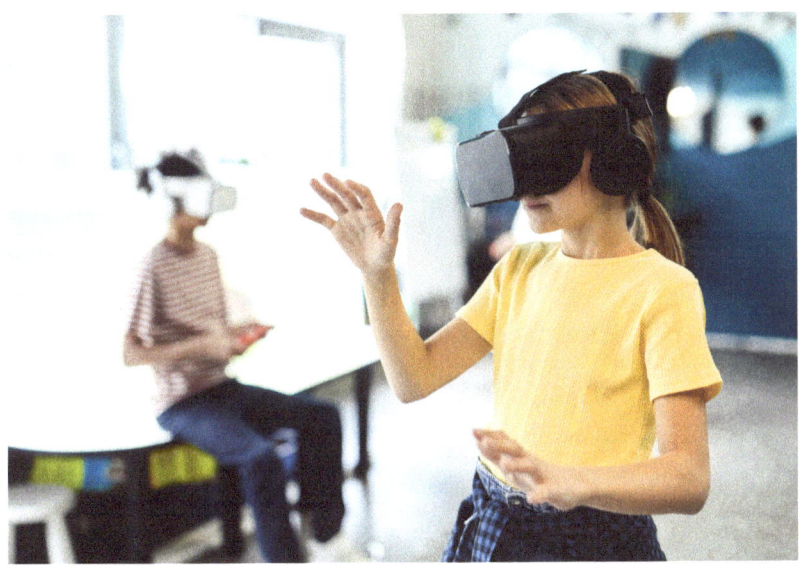

Young student using VR Set in Class
Vanessa Loring

There are a few ways to create content for VR. One is to use a software program to create the VR experience. Another is to use 360-degree video footage or photos. And finally, you can create 3D models. All of these methods have their own advantages and disadvantages. Creating VR experiences with software can be very complex, but allows you to create custom experiences tailored to your needs. Using 360-degree video or photos is simpler, but the results may not be as immersive. Creating 3D models can be time-consuming, but can yield the most realistic results. Ultimately, it depends on your needs and what you want to achieve with VR.

To get started, you'll need a VR headset and some software or 360-degree videos/photos. There are a number of different headsets on the market, and most of them are compatible with a variety of devices. The best way to find out which headset is right for you is to do some research online or go to a store and try them out. For example, the Oculus Rift is compatible with a variety of devices such as the iPhone, Samsung Galaxy, and PC. The HTC Vive is also compatible with a variety of devices such as the iPhone, Samsung Galaxy, and PC.

There are a number of software programs that can be used to create VR content. Some popular ones include Unity, Unreal Engine, and vrStorm. Unity is a popular game development software that can be used to create VR content. It offers features such as a built-in physics engine, asset import, and shader editor that can help you create high-quality VR experiences. Unreal Engine is another popular game development software that can be used to create VR content. It offers features such as dynamic lighting, particle effects, and post-processing that can help you create immersive VR experiences. vrStorm is a VR development tool that allows you to create 3D VR experiences using prefabricated 3D assets or your own 3D models. A 360-degree camera can be used to capture video and photos to be viewed in virtual reality.

Before you get started, determine if you would like to create virtual reality content by using 3D modeling software, 360-degree video filming, or photogrammetry. 3D modeling software is used to create three-dimensional models of objects or scenes, which can then be used in virtual reality environments. 360-degree video filming captures a scene in all directions simultaneously, allowing viewers to rotate their view to see different parts of the scene. Photogrammetry uses photographs to create three-dimensional models of objects. Once you determine the best method for you, find the necessary tools and equipment, and get just started.

| six |

Benefits & the Future of VR

Virtual reality has many different applications from gaming and entertainment to education and business. By using a virtual reality headset, you can immerse yourself in a new world. This is a great way to experience new and exciting environments. People will experience virtual reality in a number of ways over the next few years. There are a few ways to experience virtual reality without a headset. Some glasses, like Google Cardboard, can be used to view virtual reality content on your phone. There are also some websites that allow you to view virtual reality content without a headset.

Senior Man using Virtual Reality Glasses
Kampus

Some will use head-mounted displays such as the Oculus Rift, Oculus Quest, HTC Vive, or Sony PlayStation VR. Others may use devices that are inserted into their eyes like the Samsung Gear VR.

There are also standalone devices like the Google Daydream View and Lenovo Mirage Solo that do not require a phone or computer.

Regardless of the mechanism, virtual reality can be used for business, entertainment, such as video games and movies, or for educational purposes, such as learning about different cultures or environments.

In Education

Virtual reality is the next big thing in education. VR can provide an immersive experience that allows students to explore new places and learn in a more engaging way. These experiences can transport people to different worlds and allow them to experience things they couldn't otherwise. Students can travel to different parts of the world without leaving the classroom. This makes it an incredibly powerful tool for education, as it can give students a whole new level of understanding and engagement with the material they are learning.

> "The future is already here - it's just not very evenly distributed." - William Gibson

Thanks to recent technological developments, we're now able to use virtual reality in classrooms around the world. Teachers can use virtual reality headsets and software to create a 3D learning environment for students. This immersive educational experience has several benefits that make it ideal for certain subjects. When students are immersed into an unfamiliar environment, they learn through exploration and discovery. They have no preconceived notions or expectations about what they'll see, so their minds aren't biased by existing knowledge.

The next 10 years look very promising for virtual reality. With the continued advancement of technology, we can expect to see even more realistic and immersive VR experiences. This will be especially beneficial for educators, who can use VR to provide students with an immersive learning experience that they couldn't get any other way.

In Business

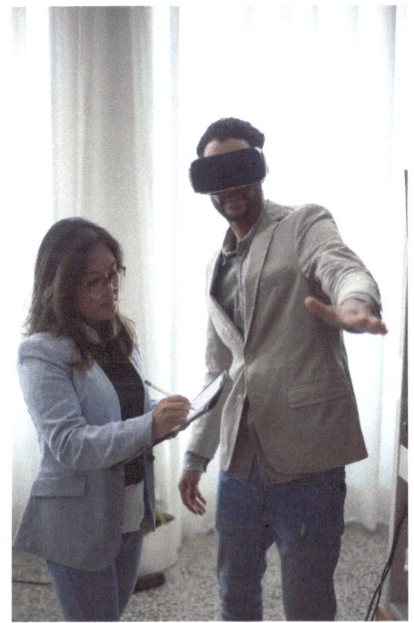

Business Man using VR Headset at work with collogue
Kampus

More and more businesses will begin using virtual reality in some way to improve their operations. Some ways businesses can use virtual reality are by having employees training in a virtual environment, testing new products in a simulated setting, and even holding meetings in a virtual conference room.

Additionally, businesses can create VR experiences to draw in customers. For example, a travel agency could create a VR experience that allows users to explore different destinations.

Utilizing VR can increase efficiency, improve customer service, and reduce costs for a business. Additionally, businesses can use virtual reality for market research, creating new products and product testing, and creating immersive customer experiences that can help boost sales and increase profit. Virtual reality technology is still in its early stages, but it has already shown a lot of potential for businesses. As the technology continues to improve, businesses should continue to explore the possibilities that VR offers to improve their operations.

In Entertainment

Virtual reality will become a bigger part of entertaining people in the future. There are countless ways VR will be used to entertain. Virtual reality will become more accessible to the public, meaning that it will be more affordable and easier to use. Secondly, there will be more virtual reality content available, so people will have more options for entertainment. Additionally, virtual reality will be more immersive, providing a more realistic experience that is more engaging.

Virtual reality is constantly evolving and changing, so it is hard to say for certain what the future of virtual reality will be. However, it is likely that virtual reality will continue to grow in popularity and become more and more widespread. Some people may question the point of virtual reality, but when used correctly, it can be an incredibly powerful tool.

WHAT EXACTLY IS VIRTUAL REALITY? — 23

Person Using Virtual Reality Technology for Gaming
Tima Miroshnichenko

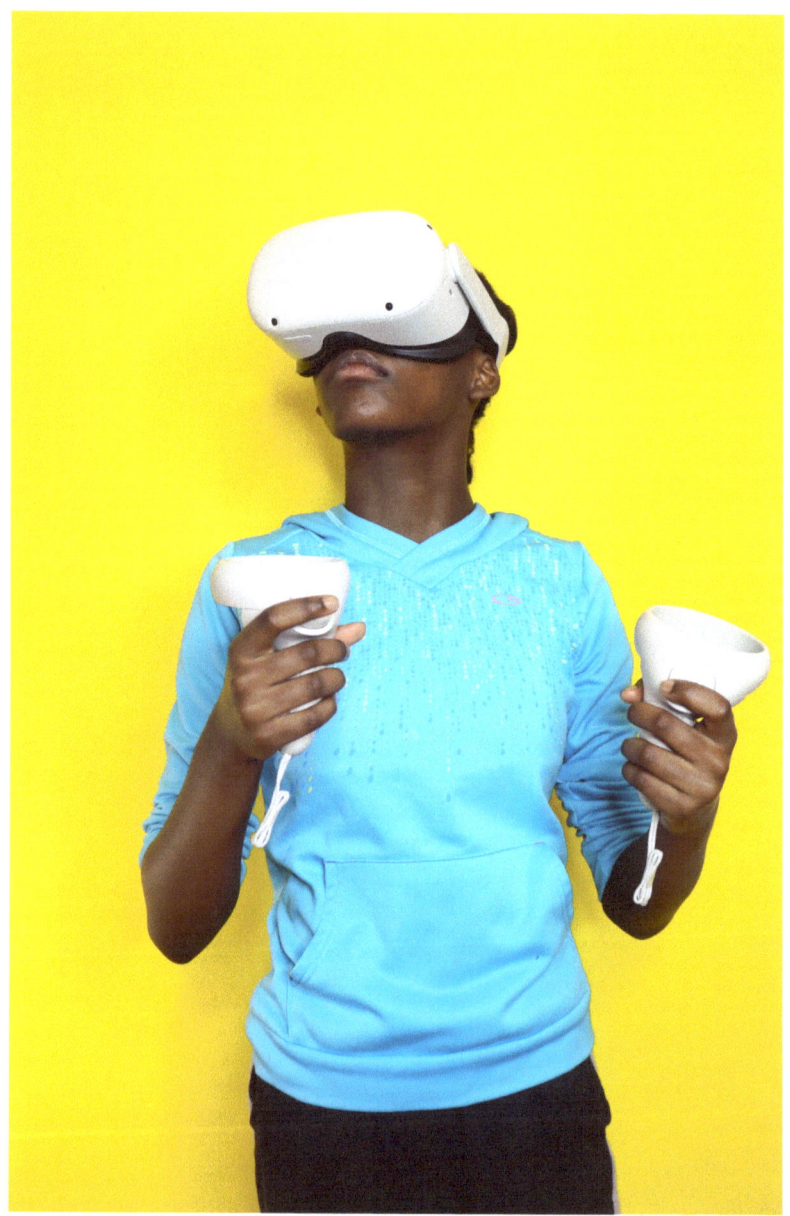

Student with Met Oculus VR Headset & Controllers
Kimberly Gordon

www.ingramcontent.com/pod-product-compliance
Lightning Source LLC
Chambersburg PA
CBHW040200100526
44590CB00006B/138